Math Monsters

NUMBER CONSERVATION
Planting Monster Melons

Based on the Math Monsters™ public television series, developed in cooperation with the National Council of Teachers of Mathematics (NCTM).

by John Burstein

Reading consultant: Susan Nations, M.Ed., author/literacy coach/consultant
Math curriculum consultants: Marti Wolfe, M.Ed., teacher/presenter; Kristi Hardi-Gilson, B.A., teacher/presenter

WEEKLY READER®
EARLY LEARNING LIBRARY

Please visit our web site at: **www.earlyliteracy.cc**
For a free color catalog describing Weekly Reader® Early Learning Library's list of high-quality books, call 1-877-445-5824 (USA) or 1-800-387-3178 (Canada).
Weekly Reader® Early Learning Library's fax: (414) 336-0164.

Library of Congress Cataloging-in-Publication Data

Burstein, John.
  Number conservation: planting monster melons / by John Burstein.
     p. cm. — (Math monsters)
  Summary: While planting two gardens, the monsters learn that the number of monster melon seeds, and the length of two hoses, does not change when the objects are moved ar rearranged.
  ISBN 0-8368-3814-9 (lib. bdg.)
  ISBN 0-8368-3829-7 (softcover)
  1. Number concept—Juvenile literature. [1. Number concept.] I. Title.
QA141.15.B86   2003
513—dc21   2003045040

This edition first published in 2004 by
Weekly Reader® Early Learning Library
330 West Olive Street, Suite 100
Milwaukee, WI 53212 USA

Text and artwork copyright © 2004 by Slim Goodbody Corp. (www.slimgoodbody.com).
This edition copyright © 2004 by Weekly Reader® Early Learning Library.

Original Math Monsters™ animation: Destiny Images
Art direction, cover design, and page layout: Tammy Gruenewald
Editor: JoAnn Early Macken

All rights reserved. No part of this book may be reproduced, stored in a retrieval system, or transmitted in any form or by any means, electronic, mechanical, photocopying, recording, or otherwise, without the prior written permission of the copyright holder.

Printed in the United States of America

1 2 3 4 5 6 7 8 9 07 06 05 04 03

You can enrich children's mathematical experience by working with them as they tackle the Corner Questions in this book. Create a special notebook for recording their mathematical ideas.

Number Conservation and Math

Exploring the concept of number conservation can help children understand that the number, quantity, and length of objects do not change when those objects are moved, rearranged, or hidden.

Meet the Math Monsters™

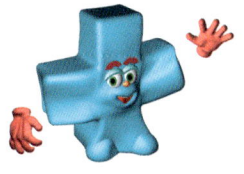

ADDISON

Addison thinks math is fun.
"I solve problems one by one."

Mina flies from here to there.
"I look for answers everywhere."

MINA

Multiplex sure loves to laugh.
"Both my heads have fun with math."

MULTIPLEX

Split is friendly as can be.
"If you need help, then count on me."

SPLIT

We're glad you want to take a look
at the story in our book.

We know that as you read, you'll see
just how helpful math can be.

Let's get started. Jump right in!
Turn the page, and let's begin!

One warm, sunny morning, Addison was outside with the other Math Monsters. He sang a happy song.

"The days are getting longer.
The sunshine feels much stronger.
It's spring! It's spring!
    We spend happy hours
    growing plants and flowers.
    It's spring! It's spring!
        We are happy for a reason.
        The reason is the season.
        It's spring! It's spring! It's spring!"

"Let's plant a garden," he said.
"We have lots of room," said Split.
"Let's plant two gardens!"

What would you plant if you had your own garden?

"Let's call Aunt Two Lips at her garden shop. We can ask her for some seeds," said Multiplex.

"Hello," said Aunt Two Lips. "May I help you?"

"Can you bring us some seeds?" asked Multiplex.

"What kind of seeds do you need? I have many kinds," she said.

What kinds of seeds do you think Aunt Two Lips has?

"If you want a special treat,
tasty as can be,
you can grow a special seed
into a pizza tree.

How about an ice cream plant
or a pretzel flower?
These seeds of mine take little time.
They bloom in just an hour."

What other kinds of pretend plants can you think of?

"Do you have any monster melon seeds?" asked Multiplex.

"I sure do," said Aunt Two Lips. "I will bring them right over."

"We are planting two gardens," said Multiplex. "Can you leave the seeds in two piles?"

"Do you want the same number of seeds in each pile?" asked Aunt Two Lips.

"Yes, please," said Multiplex.

Why do you think the monsters want the same number of seeds in each pile?

"The same number of seeds in each pile sounds fair," said Aunt Two Lips.

She drove to the garden with the seeds in her truck. She left them in two piles.

The monsters looked at one pile. They looked at the other pile.

"These piles do not look the same," said Multiplex. "How can they both have the same number of seeds?"

"Let's check," said Split.

How do you think the monsters can check?

"I will count the seeds in both piles," said Mina. She tried to count the seeds. Some seeds fell. Some seeds rolled. Some got counted two times. Some did not get counted at all.

"This is not working," said Mina.

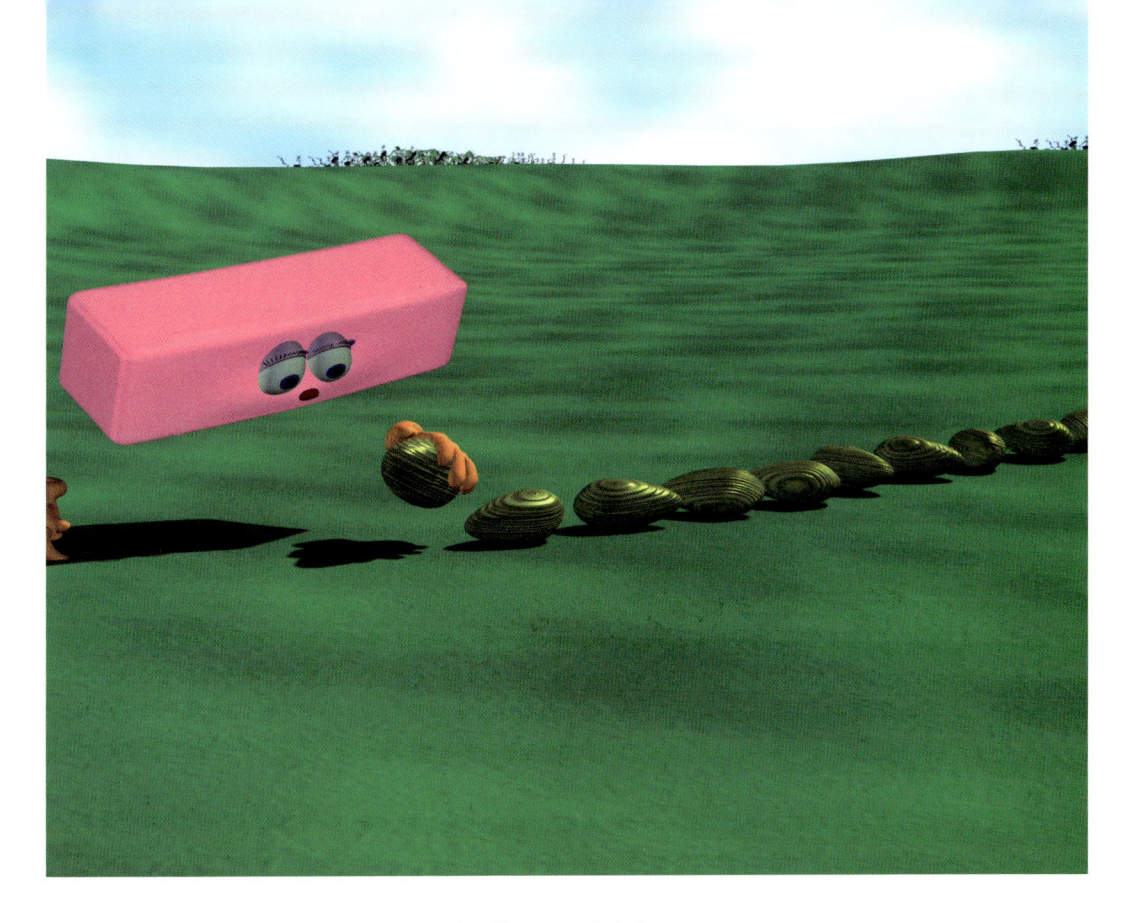

"Let's move the seeds," said Mina. "We can line them up in two rows. Then we can see which row has more seeds."

What do you think the monsters will find out?

The monsters made two rows.

"One row is short, and one row is long," said Addison.

"Aunt Two Lips must have left more seeds in one of the piles," said Split. "Let's call her and ask."

They called Aunt Two Lips. She said, "I am sure that both piles had the same number of seeds. Please count them and see."

Do you think counting may be easier with the seeds in rows? Why?

Counting the seeds was much easier in rows.
Split counted both rows.

Each row had thirty-five seeds.

"I do not understand," said Addison. "One row is longer than the other."

"How can they both have the same number of seeds?" asked Multiplex.

How can the rows look so different but have the same number of seeds?

"The seeds in one row are spread out," said Addison. "The seeds in the other row are closer together. But thirty-five seeds are the same as thirty-five seeds no matter how they look."

The monsters planted the seeds.
"Let's water them," said Split.
The monsters had two hoses.
"Is one longer, or are they the same?" asked Mina.

What do you think? How can the monsters find out?

The monsters pulled out both hoses. Split and Mina lined up the hoses side by side.

"They look the same now. They were not rolled up the same way," said Multiplex.

"Let's get to work," said Split.

Soon the seeds grew into great big monster melons. They looked yummy.

The monsters sang,

"Monster melons are a treat.
A snack like this cannot be beat.
They are so much fun to eat.
Cut in half, they taste so sweet."

Are two halves of a melon more than one whole melon or the same amount? Why do you think so?

23

ACTIVITIES

Page 5 Talk to children about all the possibilities for planting a garden. Plan a garden together. Talk about size, shape, number, and kinds of plants you would like to grow.

Page 7 Visit a garden shop with the children or look through a gardening catalog. Discuss how seeds become plants.

Page 9 Have children draw some imaginary plants.

Page 11 Making things fair and equal is an important concept for children. Discuss times when this is especially important, such as when choosing teams or being assigned chores.

Pages 13, 15, 17 Model the monsters' story using snacks such as berries, raisins, or nuts. Have the children separate one pile into two equal piles. Count them to show that they have the same number. Put the snacks into two equal rows. Count them again. Is it easier?

Page 19 Make one row longer by spreading out the snacks. Make the other row shorter by moving them closer together. Ask whether the rows still have the same number. Line up the snacks evenly to check.

Page 21 Use two strings that are equal in length. Coil one up and stretch the other one out. Ask if one is longer. Show that they are the same length. Vary the lengths of the strings with curves and zigzags.

Page 23 Roll two equal balls of clay. Have one person keep one ball whole. Have another person cut one ball in half. Discuss whether the first person has more, less, or the same amount of clay.